RACE RELATIONS CODE OF PRACTICE IN PRIMARY HEALTH CARE SERVICES

For the elimination of racial discrimination and the promotion of equal opportunities

ACKNOWLEDGEMENT

We should like to thank the many Family Health Service Authorities; Community Health Councils; Regional and District Health Authorities; General Practitioners and related health bodies and individuals throughout the country who provided us with their comments and responses to earlier drafts of this code. We are particularly grateful to the Working Group of Racial Equality Officers in London and the South for their help in preparing the document.

We thank the Secretary of State for Health, Rt. Hon. William Waldegrave MP for his endorsement of the code.

The code also reflects the experiences of people from ethnic minority communities when using the health service, as told to us. Responsibility for the code however, rests with the Commission alone.

Jackie Robinson
Senior Health and Social Services Officer

© Commission for Racial Equality
Elliot House
10-12 Allington Street
London SW1E 5EH

First published in 1992. Reprinted 1998
ISBN 1 85442 072 0
Price: £3.00

Printed by Belmont Press, Northampton

CONTENTS

	Foreword	5
1.	Introduction	7
2.	**Defining Primary Health Care**	10
	Family health services	10
	Community health services	13
	The WHO Declaration	13
3.	**The Race Relations Act 1976**	14
	Defining discrimination	14
	Health provisions of the Act	17
	Enforcement provisions of the Act	20
4	**The NHS and Community Care Act 1990**	22
5.	**Strategies for Good Practice**	25
	Race equality policies	25
	Training	29
	Positive action	32
	Language and communication	34
	Ethnic records and monitoring	38
	Recommendations for specific services	41
6.	**Racial Harassment**	43
	Bibliography	46
	Useful addresses	47

FOREWORD

I am committed to maintain a National Health Service (NHS) which is appropriate and accessible to all sections of our multi-racial community. Avoidance of racial discrimination in the provision of health services and health promotion is central to achieving this.

Primary health care services provide the vast bulk of patient care. When they are ill or want health advice it is to these services that people turn first. So, it is especially important that the services are sensitive to the needs of different racial groups and that there is genuine equity of access and provision.

It is not enough simply to agree that these must be the goals. It requires sustained commitment by all those working in primary health care, both in clinical and managerial roles throughout the primary care services, to take the necessary actions. It also means that all involved in the provision of care need to understand the law and how good practice can be developed and maintained. So, I welcome the initiative of the Commission for Racial Equality in developing its Race Relations Code of Practice in Primary Health Care Services which is designed to help with both these needs.

The NHS has already given much attention to ensuring that its policies and practices are soundly based to avoid discrimination and promote equality of opportunity. But more needs to be done. I hope that health authorities, FHSAs, NHS Trusts and practitioners will study this Code carefully, and that it will help further to inform and promote action to achieve and sustain racial equality in the primary health care services.

I look forward to the publication of the further Codes on mental health and maternity services that the Commission is preparing.

William Waldegrave
Secretary of State for Health

1. INTRODUCTION

Aims 1.1 The aims of this Code of Practice in primary health care services are:

- To provide detailed guidance on the operation of the Race Relations Act 1976 and the elimination of racial discrimination in employment[1] and service provision.

- To provide examples of good practice in the implementation and promotion of equal opportunities.

Status of the Code 1.2 The Code does not impose any legal obligations, nor is it an authoritative statement of the law; this can only be provided by the courts.

Applying the Code 1.3 The Race Relations Act 1976 applies to all National Health Service (NHS) provision and personnel including:

- Health authorities.
- NHS trusts.
- Family health services authorities (FHSAs).
- Practitioners and their staff.

---

[1] The statutory *Race Relations Code of Practice: For the elimination of racial discrimination and the promotion of equality of opportunity in employment* was produced by the Commission for Racial Equality and published in 1984 after receiving parliamentary approval.

- Community health councils.
- Other relevant voluntary organisations.

1.4 The Code does not confine itself to what is required by law, but also makes detailed recommendations. These may need to be adapted to suit individual circumstances, but any adaptations should be fully consistent with the Code's general intentions.

1.5 The Code is applicable to primary health care provision in England, Scotland and Wales.

Discrimination in primary health care services

1.6 Recent data and research studies[2] show that ethnic minority communities have suffered, and continue to suffer, from racial discrimination in the provision of primary health care services. This affects the quality, availability and accessibility of health care for ethnic minority patients and has an adverse impact on the health and well-being of ethnic minority communities in general.

1.7 In many instances, racial stereotyping by primary health care staff prevents people from ethnic minority communities from receiving the recognition and treatment to which they are entitled. One of the main criticisms of primary

---

[2] For example, *Action not Words,* National Association of Health Authorities and Trusts, 1988, and Steve Fenton, *Race Health and Welfare: Afro-Caribbean and South Asian people in Central Bristol,* Bristol University, 1984.

health care services has been that they are not sufficiently sensitive to the full range of racial, religious and linguistic diversity in Britain. Finally, the social and economic disadvantages facing ethnic minorities in Britain, compounded by discrimination, racial hostility and insensitivity, have had, and continue to have, serious consequences for their health.

2. DEFINING PRIMARY HEALTH CARE

2.1 The Department of Health[3] defines primary health care as all those health services provided outside hospital by:

- Family health services, which are administered by FHSAs, and include the four practitioner services:
  - GPs.
  - Dental practitioners.
  - Pharmacists.
  - Opticians.
- Community health services, which include:
  - Community doctors.
  - Dentists.
  - Nurses, midwives, and health visitors.
  - Other allied professions such as chiropody and physiotherapy.

Family health services

2.2 Responsibility for primary health care service provision is shared by various statutory agencies. FHSAs are responsible for managing the services provided by family doctors, dentists,

---

[3] Department of Health, 'Primary Health Care: An agenda for discussion', A consultative document, 1988.

pharmacists and opticians. FHSAs do not directly employ practitioners or their staff, but the statutory arrangements they make with practitioners – in particular those introduced in 1990 for GPs and dentists – enable them to influence significantly the development of these services.

Family doctors 2.3 A large proportion of the population is registered with family doctors or GPs, who are usually the first point of contact with the NHS. The vast majority of medical episodes are dealt with by GPs outside of hospitals, but they may decide to refer patients to a hospital, if necessary. GPs may work as part of a primary health care team including nurses, physiotherapists, chiropodists, and even social workers.

2.4 Under the NHS and Community Care Act 1990, larger GP practices may apply for budgets for their practices, to buy in a defined range of secondary services on behalf of their patients. GP practices within the scheme will receive their budgets directly from the relevant regional health authority, but the FHSA will be responsible for monitoring their expenditure against the budget.

2.5 Under the NHS and Community Care Act 1990, the FHSA will also continue to manage and oversee GP practices in the following areas:

- Cash-limited funds for GP practice staff and premises.

- Indicative prescribing budgets.
- Medical audits.
- Information technology, to help monitor prescribing and referral rates.

Dental services 2.6 Dental practitioners working within the NHS determine the number of hours they are prepared to devote to the NHS and the range of services they are willing to provide. The FHSA's contract with dental practitioners includes a requirement to explain the treatment plan to patients and its probable cost.

Pharmaceutical services 2.7 Community pharmacists working within the NHS vary in the contributions they make to family health services. Some pharmacists not only dispense medicines but also advise residential homes on the safe keeping and correct administration of drugs. Some keep records of drugs supplied to certain patients on long-term medication. Pharmacists also provide leaflets and information on health care matters.

Ophthalmic services 2.8 The FHSA is responsible for administering arrangements for opticians practising within the NHS. In April 1985, however, provision of spectacles under the NHS ceased, except for children and people on low incomes. In July 1986, even this service was terminated and replaced with a voucher system. Furthermore, the entitlement to free eyesight tests was restricted to children, young people in full-time education, people on low incomes, and the registered blind and partially sighted.

Community health services 2.8 District health authorities are responsible for assessing the health needs of their resident populations, and these are provided for by community doctors, dentists and nurses, midwives, health visitors, and members of other allied professions.

Community Health Councils 2.9 Community Health Councils are independent bodies representing the interests of patients. Often, this responsibility is fulfilled in collaboration with other statutory agencies; for example, by working with the relevant health authority to ensure that adequate cervical cancer screening programmes are available in particular localities.

The WHO Declaration 2.10 At a conference held by the World Health Organisation (WHO) in 1978, a Declaration of Alma-Ata, to which the United Kingdom is a signatory, was agreed. The Declaration included the following statements:

> Primary health care is essential health care based on practical, scientifically sound, and socially acceptable methods and technology made universally accessible to individuals and families in the community ...

> It is the first level of contact of individuals, the family, and the community with the national health system bringing health care as close as possible to where people live and work ...

> The Conference strongly reaffirms that health, which is a state of complete physical, mental and social wellbeing, and not merely the absence of disease or infirmity, is a fundamental human right ...

3. THE RACE RELATIONS ACT 1976

3.1 Several provisions of the Race Relations Act 1976 are relevant to the area of primary health care. The sections of the Act dealing with employment are important, because it is unlikely that there will be equal opportunity in service provision without equal opportunity in employment. The Commission's *Race Relations Code of Practice* in employment, has been widely distributed throughout the NHS, which is one of the largest employers of people from ethnic minorities. This Code should therefore be read in conjunction with the employment Code as well as other health service Codes, on maternity and mental health services, to be issued shortly by the Commission.

3.2 Brief descriptions of the relevant sections of the Act are given below. It should be noted that, while most of the examples of racial discrimination have been taken from real cases reported by health workers and others (and denoted by a ★), some are hypothetical.

Defining discrimination

Section 3(1) 3.3 It is unlawful under the Race Relations Act 1976 to discriminate on racial grounds, either directly or indirectly. Racial grounds are grounds of race, colour, nationality, and ethnic or national origins, and groups defined by reference to these grounds are referred to as racial groups.

Direct discrimination
Section 1(1)(a)

3.4 Direct discrimination is defined as treating a person less favourably than another on grounds of race and includes the segregation of people on racial grounds.

★ EXAMPLE

3.5 A receptionist in a dental practice tells black patients that there are no appointments available. She does so only on the basis of their colour.

★ EXAMPLE

3.6 A family doctor in a joint practice refers all black patients to the Indian doctor at the practice, solely because of their colour and not because of their particular needs.

Indirect discrimination
Section 1(1)(b)

3.7 Indirect racial discrimination consists of applying, in the circumstances covered by the Act, a requirement or condition which, although applied equally to people from all racial groups, is such that a considerably smaller proportion from a particular racial group can comply with it than others; which cannot be shown to be justifiable irrespective of the colour, race, nationality, or ethnic or national origins of the person to whom it is applied; and which is to the detriment of the person concerned because she or he cannot comply with it.

3.8 'Justifiability' means that there should be an objective balance between the discriminatory effect of the requirement or condition and the reasonable needs of the party applying the

15

requirement or condition. The needs must be objectively justified, and it is not sufficient simply to assert that the needs exist and that they are considered to be reasonable.

EXAMPLE 3.9 A community health clinic produces information leaflets in English about breast cancer and screening. The leaflets are distributed in an area where a substantial proportion of the population are not fluent in English. The health clinic refuses to translate the leaflets into the most commonly used languages in that area. In effect, this means that patients are required to read and understand English if they are to benefit from the advice contained in the leaflets.

Victimisation
Section 2(1)

3.10 It is unlawful under the Act to discriminate by victimising someone. A person is victimised if he or she is given less favourable treatment than others in the same circumstances because it is suspected or known that he or she has brought proceedings under the Act, or given evidence or information relating to such proceedings, or alleged that discrimination has occurred.

EXAMPLE 3.11 A nurse in a health centre gives evidence in proceedings brought under the Race Relations Act against a doctor who has been accused of racial discrimination towards black patients. The nurse is subsequently dismissed on unsubstantiated grounds.

Instructions or pressure to discriminate
Sections 30 & 31

3.12 It is unlawful for a person to instruct or attempt to put pressure on someone to contravene the Race Relations Act. Such pressure need not be applied directly; it is unlawful even if it is applied in such a way that the other person is likely to hear about it.

★ EXAMPLE

3.13 An optician instructs his receptionist not to register any more patients from a particular locality because of the large proportion of Africans living there.

Section 32

3.14 Liability for discriminatory acts rests both with the person who does the discriminating as well as the person who gives the instruction or applies the pressure.

Health provisions of the Act

Section 20

3.15 It is unlawful for anyone concerned with the provision of primary health care services to discriminate on racial grounds by refusing or deliberately omitting to provide the services; or as regards their quality; or the manner in which, or the terms on which, they are provided.

★ EXAMPLE

3.16 Afro-Caribbean people presenting symptoms of cancer are only prescribed aspirin because the GP does not take the time to reach a proper diagnosis.

Section 35

3.17 It is **lawful** to provide access to facilities or services to people from a particular racial group in order to meet their needs in respect of education, training, welfare, or access to any other ancillary benefits.

★ EXAMPLE 3.18 A district health authority provides resources for a centre to provide information, counselling and screening on sickle-cell anaemia and thalassaemia for the African, Caribbean, Asian and Mediterranean communities.

★ EXAMPLE 3.19 A retail pharmacist displays information about certain drugs in Gujerati and Urdu.

Employment
Section 5(2)(d)

3.20 It is **lawful** to appoint someone from a particular racial group where the job involves providing people from that group with personal services to promote their welfare, and where those services can most effectively be provided by someone from that racial group.

★ EXAMPLE 3.21 A district health authority employs an African woman to provide counselling and information on HIV/AIDS to other African women in a community health clinic. A large proportion of these women do not understand English well and, moreover, might find it difficult, or feel reluctant, to discuss sexual practices with a male worker. This service would, therefore, be provided most effectively by an African woman. It is important, however, with all section 5(2)(d) posts, to demonstrate that the appointment represents more than just a preference and that it will fulfill a need.

Section 32 3.22 An employer is liable for any discriminatory act performed by an employee in the course of his or her employment, even if it was done without the employer's knowledge or consent – unless the employer took all reasonably practicable steps to prevent such discrimination.

EXAMPLE 3.23 Staff at a community health clinic deliberately omit to give full advice and counselling to Asian women on the full range of contraception available. The clinic's management would be liable for the actions of its staff, even if it could claim to have been unaware of what had happened. However, if the management can show that it took all reasonable and practical steps to prevent discrimination, it would be relieved of liability.

Advertisements
Section 29

3.24 It is unlawful to publish, or cause to be published, an advertisement which indicates, or might reasonably be understood as indicating, an intention to perform an act of discrimination, whether that act be lawful or unlawful. (For exceptions, see the Race Relations Act 1976.)

EXAMPLE 3.25 A notice distributed among GPs in a particular area advises doctors to check the passports of all new ethnic minority patients, to ascertain their right of abode and, thereby, their entitlement to treatment.

Enforcement provisions of the Act

Commission for Racial Equality

Sections 48, 58

3.26 The Commission may conduct formal investigations into any matter, and where it discovers conduct that contravenes the Act it is empowered to issue a non-discrimination notice requiring the person or organisation on whom it is served not to contravene specified provisions of the Act. In addition, the notice can require the person or organisation to take a number of steps in order to avoid repeating such discrimination in the future. The notice may also require the person or organisation to provide the Commission with specified information, in a particular form and at particular times, in order to monitor compliance with the notice. The Commission has the power to monitor the notice for a period of five years. The person or organisation may appeal against

Section 59 any requirement of the notice, under section 59 of the Act.

Section 62 3.27 The Commission may, in certain circumstances, institute legal proceedings in respect of persistent discrimination.

Section 63 3.28 The Commission has the sole right to institute legal proceedings in respect of discriminatory practices, advertisements, and instructions and pressure to discriminate.

Section 66 3.29 The Commission may also, in certain circumstances, assist individual complainants in taking their cases to tribunals or county courts.

Rights of individuals

Section 54

Section 57

3.30 If someone believes that they have been a victim of racial discrimination, they have a right to take proceedings, in an industrial tribunal for complaints in the area of employment, and in a designated county court (in England and Wales) or a sheriff court (in Scotland) for complaints in all other areas, including primary health care.

Legal remedies

3.31 Remedies available to complainants include:

- A declaration of the rights of both parties in the matter of the complaint.

- Compensation or damages, including a sum for injury to feelings.

- An injunction or order against the discriminator.

4. THE NHS AND COMMUNITY CARE ACT 1990

4.1 The changes introduced as a result of the White Papers, *Promoting Better Health* and *Working For Patients,* and the NHS and Community Care Act 1990 have had a significant impact on primary health care services:

- FHSAs are now accountable to their regions rather than directly to the Department of Health.

- Health authorities have become purchasers of community health services via contracts with NHS trusts, Directly Managed Units, and the private or voluntary sectors.

- Larger GP practices are eligible to apply for budgets in order to purchase a range of in- and out-patient hospital services directly rather than through the district health authorities.

Areas of concern

4.2 There are three major areas of concern in respect of the NHS and Community Care Act 1990 and the provision of health services to ethnic minority communities[4]:

Language and communication

4.3 It is vital that purchasers and providers of primary health care services give proper consideration to the translation of all relevant information into the various languages used in their area and the provision of interpreting services, wherever necessary. Equally important is the matter of choosing the most appropriate channels of communication.

Consultation and participation

4.4 Purchasers and providers of primary health care services will need to plan consultations specifically with their local ethnic minority communities, to ensure that their views and participation in the planning process are secured. The arrangements should take into consideration the different languages spoken, the form of the consultation, and the venue.

Ethnic minority services

4.5 Purchasers and providers of primary health care services cannot be sure that services are free from racial discrimination and that they are appropriate, adequate and accessible to people from ethnic minority communities unless they consult them fully about the services that should

---

[4] See also the following publications and papers by the Commission for Racial Equality: *NHS Contracts and Racial Equality;* 'Response to White Paper on the Review of the National Health Service', 'Recommendations to the NHS and Community Care Bill'; and 'Response to Caring for People: Draft Guidance Circulars'.

be provided, assess the services, and evaluate their effectiveness jointly with the local communities.

4.6   General practitioners are now working under new contracts, and first-wave GP fundholders are using their budgets effectively. The FHSAs are monitoring both these changes in primary care. While it is true that single-handed practitioners find it more difficult to provide a comprehensive range of services, this has little to do with budget-holding. The underlying causes tend to be professional isolation, proportionately higher costs, and smaller surgeries

4.7   Providers of primary health care services who have contractual agreements with the health authority and the FHSA should be advised to demonstrate that they have considered the needs of people from ethnic minority communities and included anti-discriminatory measures in their procedures and policies.

4.8   One way of evaluating the effectiveness of a contractor's race equality record is for health authorities and FHSAs to include performance indicators of the contractor's race equality practices as part of the information required for inclusion in any approved list of contractors, or in the actual terms of contracts.

5. STRATEGIES FOR GOOD PRACTICE

Race equality policies

5.1 The adoption and implementation of race equality policies in the purchasing, provision and administration of primary health care services are essential if ethnic minority communities are to have confidence that racism is being challenged, and discrimination eliminated throughout the service. Race equality policies will also go some way towards ensuring that adequate, appropriate and accessible services that meet the needs of all sections of the community are provided.

Basic principles

5.2 The basic components of a race equality policy are:

- A statement of the organisation's commitment to the elimination of racial discrimination and the administration, development and provision of services that are adequate, appropriate and accessible to all sections of the community.

- A brief summary of the actions that will be taken to achieve this objective throughout the organisation and in all the services it provides.

5.3 Larger practices, community health centres and clinics may wish to develop their own race equality policies, in conjunction with the FHSA or health authority. Smaller practices should ensure that they are aware of the FHSA's or health authority's race equality policy, and seek guidance on how to implement and sustain anti-racist practices in their services.

5.4 Community Health Councils, public information bodies, and other relevant voluntary organisations should also adopt race equality policies and implement them according to local circumstances and needs.

5.5 Ethnic monitoring of service users and regular reviews of policies and procedures will disclose any differences in rates of participation, success or progress between ethnic groups, and provide the basis for action to remove barriers. Without a monitoring system, no purchaser or provider of primary health care services can be sure that it is using its resources fully and providing services that are adequate, appropriate and accessible to all patients.

5.6 The race equality policy should include:

i A declaration that the purchaser or provider will abide by the Race Relations Act 1976 and implement the provisions of the Race Relations Code of Practice in Primary Health Care Services as far as practicable.

ii An undertaking to notify employees, patients and, where applicable, practitioners, of the adoption of the policy (translated where appropriate), and to make the Race Relations Code of Practice in Primary Health Care Services available for inspection.

iii An undertaking to allocate responsibility for implementing the overall and individual parts of the policy at a senior level, and to identify clear lines of accountability, so that progress can be properly monitored and results reported at regular intervals to managers or appropriate authority members.

iv The establishment, where possible, of a Race Equality Committee, comprising representatives from the FHSA or health authority, individual practitioners or representatives of groups of practitioners, and members of ethnic minority communities.

Implementing the policy 5.7 All purchasers and providers of primary health care services who adopt race equality policies should develop an action programme incorporating the following steps.

Reviews 5.8 The practices and procedures of the health authority, the FHSA and practitioners, and the primary health services they are responsible for should be regularly reviewed to ensure that they do not discriminate directly or indirectly, and changes introduced where it is found that they are, or may be, contravening the Race Relations Act.

Code of Practice in Employment

5.9 The Race Relations Code of Practice in Employment should be adopted by the health authority, the FHSA, and primary health care providers, and progress on its implementation regularly monitored. Consideration should also be given to implementing the positive action provisions of the Act.

Race equality training

5.10 Once a race equality policy has been adopted, all staff should receive training and guidance on it, to ensure that they understand clearly the legal position and the implications of the organisation's policy. (See also paras. 5.14 – 5.21.)

Allocating resources

5.11 It is essential that the different needs of ethnic minority patients are recognised as valid and essential factors when determining levels of funding. It is recommended that financial planning for race equality initiatives should be an integral part of policy development.

Ethnic records and monitoring

5.12 The race equality policy should be fully monitored. One way of doing this is to keep ethnic records and monitor them on a regular basis. The Department of Health recognises the importance of ethnic monitoring in the provision of health services. The Commission recommends that health authorities and FHSAs should keep ethnic records in terms of the services they purchase and provide, to ensure that they do not discriminate. Where practicable, practitioners should be encouraged to do the same. (See also paras. 5.39 – 5.41.)

Unmet needs 5.13 Any unmet needs, such as screening for specific illnesses, or interpreting and translation services, should be identified, and action taken to meet them.

Training

5.14 Training is vital for the effective implementation of non-discriminatory practices and procedures, not only to ensure that staff understand them, and their own specific responsibilities, but also so that they can learn how to change their ways. However, training is only one of several tools in helping to provide adequate and appropriate services to ethnic minority communities; it should, therefore, be seen as part and parcel of a wider programme of race equality strategies.

Consultation 5.15 Arrangements should be made to consult the ethnic minority communities and ensure their participation in the development, content and delivery of training programmes.

Funding 5.16 Race equality training programmes should be funded directly out of mainstream budgets, so that they are not marginalised or lose credibility.

Types of training 5.17 Several types of training will be needed to equip staff at various levels for their respective responsibilities in eliminating racial discrimination. In general, though, all training should have an anti-racist perspective, with examples of racial discrimination taken from primary health care services, so that staff can relate to the issues. Sometimes, it may be necessary to 'import' expertise. It is important that the training sessions include time for informal discussion, so that the 'hows and whys' can be fully considered by those whose support will be crucial.

Basic training

5.18 All basic training programmes, including programmes for staff involved in reception and administrative duties, should provide:

- Examples of racial discrimination that are relevant to the particular service provided.

- Awareness of cultural differences and their effects on health patterns and patients' needs.

It is recommended that representatives of ethnic minority groups are invited on a regular basis to participate in basic training programmes. Care should be taken, though, to see that their involvement is appropriate, constructive, and useful to them, and not just a token exercise.

5.19 Staff directly involved in patient care should also be trained to recognise, and avoid, racially discriminatory treatment. It is essential that ethnic minority patients are not insensitively and inappropriately categorised on the basis of preconceived cultural stereotypes about certain ethnic groups. While recognising their particular needs, the training should direct staff to relate to ethnic minority patients as individuals, and devise strategies for ensuring that no hostility is generated and that patient and carer communicate on equal terms.

Management training

5.20 Senior managers must understand fully why race equality initiatives are needed, and their specific objectives. Their training should cover:

- The legal framework, including the Race Relations Act 1976, thus providing a wider basis for their understanding of, and commitment to, non-discriminatory services.

- The operational details of all race equality policies and initiatives.

- Leadership skills, to direct and inspire other staff.

Race equality training should also be seen as an integral part of the professional competencies for the various disciplines associated with medicine, and not marginalised as an optional extra.

In-service training

5.21 In-service training should take account of both the needs of the communities being served and the ethnic backgrounds of individual staff members. An understanding of racial discrimination and its effects, and a broad appreciation of the religious, linguistic and cultural needs of ethnic minorities should be an important part of any in-service training programme.

Positive action

Section 37

5.22 Section 37 of the Race Relations Act 1976 allows any person to provide members of a particular racial group with access to facilities for training for particular work, or encouragement to 'take advantage of opportunities' for doing that work when they have been underrepresented in that work at any time within the previous twelve months.

5.23 For the purposes of section 37, a racial group is underrepresented in particular work where it appears to the body that, at any time within the previous twelve months, there was no-one of that racial group doing that work in Britain, or the proportion of people of that racial group doing that work was small in comparison with its proportion of the population of Great Britain as a whole. Where the conditions are met only for a particular area, training or encouragement can only be provided for those who appear likely to take up that work in that area.

5.24 Section 37 does not allow employers to discriminate when recruiting for posts in their organisation. Trainees cannot be guaranteed jobs by virtue of their successful completion of a positive action training course. Section 37 only enables someone to provide appropriate training for individuals from a particular racial group in order to equip them to compete on the job market on the same basis as other racial groups, because they have been denied those skills through past discrimination.

★ EXAMPLE 5.25 The proportion of Afro-Caribbean health visitors in the area covered by a health authority is small. The authority therefore encourages Afro-Caribbean nurses to apply for training as health visitors.

Section 38(1)(a) 5.26 Section 38 (1)(a) of the Race Relations Act 1976 permits employers to select members of a particular racial group from their employees and train them to do jobs in which, over the previous twelve months, members of that group have not been represented at all, or have been underrepresented. A group will be underrepresented if there are no persons of that racial group doing the job at that establishment, or their numbers are low among the workforce as a whole in comparison to their numbers among the population of the area from which the organisation recruits.

Section 38(1)(b) 5.27 Section 38(1)(b) also allows employers to 'encourage' persons of particular racial groups to 'take advantage of opportunities' for work with that employer. This could mean applying for jobs or for training placements.

EXAMPLE 5.28 A district health authority encourages Chinese and Vietnamese people to apply for positions as link workers.

5.29 Section 38 cannot be applied unless there is underrepresentation. It is, therefore, essential that an employer can prove that there is underrepresentation of particular racial groups in his or her employ, or in the wider context of the area from which the employer would recruit people.

Language and communication

5.30 Language and communication differences have been the most notable barrier to getting health information across. The translation of information material and the provision of qualified interpreters will be vital in some areas, but it is also important that both professionals and patients recognise the cultural determinants of communication and the different ways in which people from different backgrounds behave and respond. The National Association of Health Authorities and Trusts (NAHAT) says in its report, *Action Not Words* (1988) that:

> ... language differences should not be perceived to be the patient's problem, but an issue to be overcome by the providers of the health service.

5.31 Interpreting and translating are complex and undervalued skills. Health authorities and other providers of primary health care services must recognise the importance of the contribution interpreters make to the appropriate provision of services. However, they must also remember that providing services that take account of ethnic minority needs involves more than just overcoming language barriers.

Language facilities

5.32   As far back as 1985, the London Interpreting Project (LIP) had pointed out that:

> ...large sections of the non-English speaking communities are unable to benefit from basic services.... health is perhaps the greatest area of need for interpreting services. The anxieties people face when they are ill are enormous, and they are particularly acute for people whose first language isn't English.

In 1989, in its *Directory of Interpreting Services in the Greater London Area*, LIP found that health authorities had been slow to recognise the need for interpreters as a serious and valid need. It concluded that, even where paid interpreting facilities did exist, the majority were poorly resourced, with no recognised pay or career structures, and the role of the interpreter was unclear and unsupported.

5.33   In a large number of health authorities, the need for interpreters is still only acknowledged by having lists of in-house bilingual staff who can be called upon to interpret. This is unsatisfactory in many ways:

- The ability to speak two languages is not the same as having professional interpreting skills.

- There is no assessment of the level of the bilingual person's language ability or knowledge and understanding of medical terms.

- It puts bilingual staff in an awkward position, as they may feel obliged to help, at the expense of their own work and without reward.

5.34 Another fairly common practice is to encourage patients to bring their own unpaid and, often, unskilled interpreters – relatives, friends, even their own children. This method too is unsatisfactory:

- It does not provide professional interpreting.
- It may cause embarrassment and distress both to the patient and to the interpreter.

5.35 The development of clear job descriptions and career structures for interpreters in health authorities is clearly linked with the need for professional training. Few health authorities have made arrangements for their interpreters to attend training courses to improve their language skills, medical knowledge, or understanding of the health authority itself.

5.36 LIP produces a directory of community interpreting services and resources in the Greater London Area. It can also give advice about interpreting services nationally, and provide training for those in the field or those wishing to set up community interpreting services in other parts of the country.

5.37 The Ethnic Study Group of the Coordinating Centre for Community and Health Care, has produced a *Code of Practice for*

Interpreters. We recommend that both purchasers and providers of primary health care services obtain copies and distribute the information to all staff.

Overcoming communication barriers

5.38 Professionals and patients bring their own expectations to any interaction. Primary health care staff may have negative feelings and expectations towards ethnic minority patients, and ethnic minority patients may feel awkward, under stress, or embarrassed. We offer suggestions below of how some of these barriers to communication might be tackled.

i   The adoption of a separate policy which covers guidance on localities.

ii The employment of trained interpreters in those community languages where there is a large demand for services.

iii Training for staff on how to make the best use of interpreters.

iv Displaying signs and notices in English as well as the most commonly used community languages.

v   The production of leaflets, posters, and other publicity material that reflect the multi-racial composition of the population and use positive images of people from ethnic minority communities, both as providers and receivers of services, so that all groups feel that the information applies to them.

vi The publication of well-translated leaflets

on important health topics and topics of special relevance to people who are unfamiliar with primary health care service provision; for example: how it works, how to get the help you need, how to choose and change your GP, and patients' rights.

vii Training programmes that teach staff to improve their communication skills, examine their expectations of ethnic minority patients, and give them the knowledge and awareness needed to work effectively with patients from all ethnic groups.

viii The development of mainstream career structures for interpreters.

Ethnic records and monitoring

5.39 It is recommended that purchasers of primary health care services should keep and monitor ethnic records as a way of knowing how their assessment of health service needs and the objectives they set themselves translate into actual service provision and delivery. Health authorities and FHSAs should encourage health service providers to do the same.

Basic elements of a monitoring system

5.40 The main components of an effective monitoring system are:

i An understanding that ethnic monitoring of services is an integral part of a general race equality strategy which also includes employment matters, training programmes, reviews of policies and procedures, etc.

ii   Consultation with the local communities, to explain why monitoring is being introduced, and to obtain their views on questions such as the ethnic classification system[5].

iii   The confidentiality of individual records.

iv   The designation of senior officers with overall responsibility for introducing and maintaining the system.

v   Monitoring ethnic records on a regular basis, with analytical reports produced at least every six months. The main areas that should be monitored are:

- Access – Who is receiving which services and by what process? (This will include how the public is informed about services.)

- Adequacy – Are the services provided adequate to the needs of those using them?

- Quality – What types of services are being provided, and do they meet the needs of those using them?

Ethnic records can also be used to assess the situation of particular groups in terms of the services provided; for example, ethnic

---

[5] The Commission recommends a classification system based on the one used by the Office of Population Censuses and Surveys (OPCS) for the 1991 census.

minority women and children, or the elderly; and to ensure that any particular needs they might have are being met, and that they are being treated fairly and equally.

vi Making monitoring reports, along with medical records, available to patients.

vii If a monitoring report reveals particular problems, such as low take-up by ethnic minority women of, say, cervical and breast screening services, a suitably timetabled programme of action should be agreed to deal with them.

viii Establishing an agreed complaints procedure for patients, and seeing that all patients and workers know about it.

Targets 5.41 In order to assess whether different ethnic groups are being treated equally, a standard of measurement, or target, needs to be agreed of the proportions of ethnic minorities that could be expected to use a particular service in the absence of any discriminatory barriers. The target should be based on the size of the local ethnic minority population as well as other indicators of need, such as the number of ethnic minority children below school age. Deciding on a suitable target can be a complex task initially, as it involves taking account of the size, age structure, and needs of the ethnic minority population, by ethnic group. Consultation with Racial Equality Councils and the relevant voluntary and statutory health organisations may be helpful in setting initial targets, but these should be kept under regular review. The introduction of an ethnic question in the 1991 Census will also give providers of primary health care services a valuable ethnic profile of their local populations[6].

5.42 The target will provide a yardstick by which the monitoring data are analysed to check whether ethnic minorities are receiving their fair share of services. It should be emphasised that targets must not be confused with predetermined quotas, which are unlawful under the Race Relations Act.

Recommendations for specific services

Pharmaceutical services

5.43 Problems of inadequate communication between ethnic minority communities and health service workers may be particularly pronounced in the pharmaceutical services. Ethnic minority customers with little or no knowledge of English may be unable to understand vital information about drug treatment, allergic reactions, and the side effects of medication. We recommend that:

- Pharmacists should use and display information in a form that is readily

---

[6] See also the Commission's recently published guides, *NHS Contracts and Racial Equality: A guide*; *A Measure of Equality: Monitoring and achieving racial equality in employment*; and *Accounting for Equality: A handbook on ethnic monitoring in housing*.

accessible to members of ethnic minority groups, using the relevant community languages where necessary.

- Instructions on the dosage and use of drugs should be translated into the appropriate community languages.

Ophthalmic and dental services

5.44 There is some concern about testing and routine preventive checks and recall for ophthalmic and dental services, particularly in the case of children and the elderly in ethnic minority communities. We recommend the following steps:

- Campaigns to encourage regular attendance should take account of the different languages spoken in an area and publish translations of all relevant information, where necessary.

- Ethnic minority communities should be made aware that certain conditions to which they may be more susceptible, such as hypertension and diabetes, may be identified through ophthalmic and dental checks.

6. RACIAL HARASSMENT

6.1 Racial harassment is a serious problem and does occur in the primary health care sector. Racial harassment includes not only physical attacks on people but also verbal abuse and any other form of behaviour that deters people from using primary health care services. Primary health care service providers who do not deal effectively and fairly with racial harassment may, in some circumstances, be contravening section 20 of the Race Relations Act 1976.

6.2 Similarly, GPs and other primary health care service providers should be alert to the possibility that stress and certain other conditions may be caused by racial harassment.

6.3 Community Health Councils, FHSAs and health authorities can play an important role by providing advice and guidance to health service providers on identifying and monitoring racial harassment and taking appropriate action.

Countering racial harassment

6.4 Both patients and staff may experience racial harassment; patients may be harassed by other patients or by staff, and staff by patients or by other members of staff. We recommend the following steps to deal with this problem:

i It is important to identify the perpetrators, both because of the seriousness of the offence, and because it shows that racial harassment will not be tolerated.

ii Cases should be monitored carefully, and appropriate case notes and witness statements prepared on individual incidents.

iii In certain circumstances, the police should be contacted.

iv Disciplinary action should be taken if the perpetrator is a member of staff. Where he or she is a patient, the sanctions might include restrictions on attendance at a particular health clinic, or removing the patient altogether from the practitioner's list.

v A proper complaints procedure should be drawn up, incorporating an anti-racial harassment strategy. The basic elements of the strategy should include:

- Providing advice and information.

- Making arrangements to liaise with the relevant agencies, such as: the Department of Health, police, law centres, and voluntary agencies.

- Producing information and publicity material, in all the relevant languages, on the help available to patients and the local community.

6.5 Practitioners should be made aware of the views of the FHSA, the Community Health Council, and the health authority, and should be informed of the action they should take in cases of racial harassment, or which may be taken against them by patients or the health authority where appropriate. Practitioners should make it clear to patients that racial harassment will not be tolerated, and that action will be taken when incidents occur.

6.6   Primary health care providers who adopt anti-racial harassment policies should ensure that channels of communication and responsibility are clearly identified and reporting systems introduced. The policies should be monitored to check that they are effective, and regular reports produced for relevant authority/committee members and/or managers on the results.

BIBLIOGRAPHY

Commission for Racial Equality (1984), *Race Relations Code of Practice: For the elimination of racial discrimination and the promotion of equality of opportunity in employment.*

Commission for Racial Equality (1990), 'Response to the White Paper on the Review of the National Health Service'.

Commission for Racial Equality (1990), 'Recommendations to the NHS and Community Care Bill'.

Commission for Racial Equality (1991), 'Response to Caring For People: Draft Guidance Circulars'.

Commission for Racial Equality (1991), *NHS Contracts and Racial Equality: A Guide.*

Department of Health (1988), *Primary Health Care: An agenda for discussion.*

Ethnic Study Group of the Coordinating Centre for Community and Health Care (1991), *Code of Practice for Interpreters.*

Fenton, Steve (1984), *Race, Health and Welfare: Afro-Caribbean and South Asian People in Central Bristol,* Bristol University.

London Interpreting Project (1989), *Directory of Interpreting Services in the Greater London Area.*

National Association of Health Authorities and Trusts (1988), *Action not Words.*

The NHS and Community Care Act 1990, HMSO.

The Race Relations Act 1976, HMSO.

USEFUL ADDRESSES

National Association of Health Authorities and Trusts
Birmingham Research Park
Vincent Drive
Birmingham
B15 2SQ

The Kings Fund Centre
126 Albert Street
London
NW1 7NF

National Health Service Training Authority
St. Bartholomews Court
18 Christmas Street
Bristol
BS1 5BT

Royal College of General Practitioners
14 Princes Gate
London
SW7 1PU

Royal College of Nursing
20 Cavendish Square
London
W1M 0AB

Association of Community Health Councils for England and Wales
30 Drayton Park
London
N5 1PK

British Medical Association
BMA House
Tavistock Square
London
WC1H 9JP

General Medical Council
44 Hallam Street
London
W1N 6AE

1991 Ethnic Switchboard
2B Lessingham Avenue
Tooting
London
SW17 8LU

Ethnic Study Group
Coordinating Centre for Community and Health Care
2B Lessingham Avenue
Tooting
London
SW17 8LU

Greater London Action For Race Equality (GLARE)
Room 312
South Bank House
Black Prince Road
London
SE1 7SJ

National Association of Racial Equality Councils
8-10 Coronet Street
London
N1 6HD

National Council for Voluntary Organisations
26 Bedford Square
London
WC1B 3HU

Overseas Doctors Association
28-32 Princess Street
Manchester
M1 4LB

The Maternity Alliance
15 Britannia Street
London
WC1X 9JP

Health Education Authority
Hamilton House
Mabledon Place
London
WC1H 9TX

Organisation for Sickle Cell Anaemia Research
22 Pellatt Grove
Wood Green
London
N22 5PL

UK Thalassaemia Society
c/o Mukesh Kotecha
107 Nightingale Lane
London
N8 7QY

COMMISSION FOR RACIAL EQUALITY

The Commission for Racial Equality was set up by the Race Relations Act 1976 with the duties of:

- Working towards the elimination of discrimination.
- Promoting equality of opportunity and good relations between persons of different racial groups.
- Keeping under review the working of the Act, and, when required by the Secretary of State or when it otherwise thinks it is necessary, drawing up and submitting to the Secretary of State proposals for amending it.

London (Head Office)
Elliot House
10-12 Allington Street
London SW1E 5EH
☎ 0171-828 7022

Birmingham
Alpha Tower (11th floor)
Suffolk Street Queensway
Birmingham B1 1TT
☎ 0121-632 4544

Leeds
Yorkshire Bank Chambers
(1st floor)
Infirmary Street
Leeds LS1 2JP
☎ 0113-243 4413

Manchester
Maybrook House (5th floor)
40 Blackfriars Street
Manchester M3 2EG
☎ 0161-831 7782

Leicester
Haymarket House (4th floor)
Haymarket Shopping Centre
Leicester LE1 3YG
☎ 0116-242 3700

Scotland
45 Hanovers Street
Edinburgh EH2 2PJ
☎ 0131-226 5186

Wales
Pearl Assurance House (14th floor)
Greyfriars Street
Cardiff CF1 3AG
☎ 01222-388977